I0605211

# INSIDE
## HIDDEN MÉRIDA AND BEYOND
# YUCATÁN

JO BAER

# INSIDE YUCATÁN
## HIDDEN MÉRIDA AND BEYOND

SUSANA ORDOVÁS
PHOTOGRAPHY GUIDO TARONI

VENDOME
NEW YORK • LONDON

# FOREWORD

The Yucatán Peninsula, with its rich tapestry of history, culture, and natural beauty, has always captivated those who venture into its embrace. However, the houses that dot this landscape hold stories often left untold—of families, traditions, and architectural wonders. In this splendid volume, Susana weaves a narrative that bridges the past and the present, showcasing the vibrant heart of the Yucatán through its homes.

Susana and I first became "friends" on Instagram, as one does. One day she proposed a Yucatán portfolio for *Cabana* magazine, which I obviously thought was very intriguing. I asked her to send through a few "scouting pictures". What came into my inbox a few days later exceeded all expectations and it is, to date, one of my favorite portfolios. This book, in a way, is an extended version of those first images, capturing the magic in the peninsula that resides in its patina and whirlwind of colors and patterns.

The brilliance of *Inside Yucatán* doesn't rest solely on its well-researched content. Guido's photographs act as windows into these houses, each frame a testament to his unparalleled skill and artistry. His lens not only captures the architectural elegance but also the innate soul of these homes, the whispers of generations gone by and the hopes of those yet to come. As with everywhere he goes, Guido captures the very essence of places, their DNA.

For those familiar with the Yucatán, this book will be a nostalgic stroll down memory lane. For others, it promises a compelling introduction. Dive in, and let the houses of the Yucatán tell you their stories.

Enjoy the journey!

Martina Mondadori

# CONTENTS

## TOWN 147

# INTRODUCTION

I have always been captivated by beauty in all its forms. Throughout my early years, I assumed that most people shared this obsession, but it was only later, in adulthood, that I realized my love for beautiful things and surroundings was not as common as I believed. From a young age, I was an avid reader; books fueled my imagination and allowed me to escape to wondrous worlds. Therefore, it feels serendipitous that one day I would journey through Yucatán and become enraptured with its remarkable homes and haunting haciendas, ultimately turning my fascination into a book.

To say I had an unconventional upbringing is an understatement. Both my parents were adventurous and carefree, and instilled in me a profound curiosity for life. For years, I yearned for a sense of stability and a bit of normalcy in my existence. Now that I look back, I have a deep appreciation for the way they brought me up—it proved a great help to me in the life I was to lead. I was born in Ireland, in the windswept, coastal town of Dún Laoghaire, to an English mother and a Spanish father. I spent my early years in Spain, and when I was barely six, we moved to Africa: first to Zaire (now the Democratic Republic of the Congo), where I spent the rest of my disrupted childhood, and then to Morocco when I was a teenager. The realm I inhabited as a girl expanded as I discovered the world—and especially Africa, a continent that, in the 1970s and 1980s, was still experiencing the aftermath of colonialism and imperialism. Living there had a profound effect on me.

I recall flying into Mexico for the first time thirty years ago. I was working for a press agency that took me to remote corners of the globe, and I was looking forward to finally working in a country where I could speak the language and relate to its people and culture. I was also excited about the prospect of visiting a place I had heard so much about. My parents had traveled here when I was a child and came back with fascinating stories of an alluring land. As I settled into my new life in Mexico City, I became enthralled by the vibrant culture and historical richness that surrounded me. The sprawling metropolis became my home, yet part of me longed for new horizons. I had read about Yucatán, the home of atmospheric haciendas and a peerless ancient civilization, and my curiosity was piqued.

My first visit to Yucatán was uneventful. I spent three short days in Mérida roaming the dusty streets of the *centro histórico* under a blistering sun, wilting in the heat while I admired the severe façades of the colonial *casonas* that line the busy avenues. True to their Moorish-Spanish ancestry, they maintained an old-world seclusion behind high walls, and I wondered what lay beyond them. It was not until many years later, after having traveled extensively throughout Mexico with my husband and two children, that I decided it was time to properly uncover the secrets of this remote region.

That trip, in the fall of 2020, turned out to be a transformative experience, a spiritual voyage that carried me far from my familiar urban life in Mexico City and introduced me to a world where ancient stories intertwined with the present. I became hopelessly enchanted with Yucatán's faded glory and charm. The crumbling haciendas and archeological ruins that adorned the lush tropical countryside, as well as the austere colonial buildings and decadent mansions that graced the streets of Mérida, revealed tales of splendid bygone eras. These remnants of an illustrious past were shrugging off the effects of a centuries-long *siesta*. It was evident that Yucatán was experiencing a new dawn.

Much like the Africa in which I was brought up, Yucatán cast its spell on me and irremediably got in my blood. Its storied history, cultural heritage, natural landscapes, and mixture of influences have combined to create a region with a singular character that sets it apart from the rest of Mexico. This Land of the Maya captivates and stirs the soul with its blend of indigenous and Spanish heritage. A vast, flat territory, lapped by the warm waters of the Gulf of Mexico on one side and the alluring Caribbean Sea on the other, this once isolated peninsula has fascinated travelers for centuries.

To understand this isolated region of Mexico, one must know its history. Breathtaking archaeological sites built by the ancient Maya adorn the lush tropical terrain. These historical places are filled with soaring pyramids, ornately decorated temples, and expansive palaces that reveal a civilization of exceptional sophistication and complexity. As John Lloyd Stephens and Frederick Catherwood discovered during their travels in the 1840s, the Maya civilization achieved remarkable architectural feats by constructing countless intricate and monumental structures on the rocky and infertile land that covered most of their territory. Take, for example, the Temple of Kulkulkan, at Chichén Itzá, renowned for its symmetry, or the Palace of the Governor in Uxmal, elevated on a platform and set against the blue sky with its intricate frieze. The Maya of Yucatán were master masons who strove to achieve a level of quality that was challenging to attain with the modest and rudimentary tools available to them. Their advanced engineering allowed for innovative constructions like corbel arches and complex water management systems. The detailed carvings, sculptures, and hieroglyphics on their buildings conveyed rich narratives of mythology, history, and power dynamics.

Spanish rule in Yucatán—a process that began in the sixteenth century—proved pivotal and transformative to the region. It marked the beginning of European influence, and had a huge impact on architecture and urbanization, leaving a lasting imprint on the area's built environment. The Spaniards brought their own cultural baggage to the native traditions that had flourished here, and introduced architectural styles characterized by arches and ornate façades, and stone masonry defined by the use of mortar. Over time, the indigenous people of Yucatán, descendants of the great builders and sculptors of the pre-Columbian era, combined their designs with decorative schemes brought from Spain (a country of diverse influences itself). The convergence of Spanish and indigenous styles during the colonial period led to the development of a distinctive architectural vocabulary. Notable examples include the construction of churches and cathedrals atop pre-existing Maya temples, and the repurposing of stones from Maya structures in the creation of colonial buildings, blending religious significance with historical continuity.

For a brief period, around the turn of the twentieth century, Mérida was said to house more millionaires than any other city in the world, thanks to the mass production of henequen, a fibrous agave plant native to Yucatán that was used to craft high-quality rope, twine and sacks to transport grain. As a result of this boom, the region underwent a period of astonishing prosperity unparalleled by anything Mexico had ever experienced. Pre-existing cattle and corn haciendas, founded during Spanish dominion, were converted into extravagant henequen plantations, and the affluent *hacendados* also built lavish homes for their families in Mérida. Paseo de Montejo, the city's main avenue, which was built to emulate the Champs-Élysées in Paris, boasts ornate mansions and is testament to the incredible wealth amassed during this time. It was this combination of historical factors that produced the vitality of the unrivaled design heritage that Yucatán possesses today, as this book so vividly explores.

In recent years, Yucatán has evolved into a haven for international artists, designers and creatives who have been drawn to its old-world architecture and dynamic cultural heritage. In fact, many acknowledge a curious sense of homecoming when they first arrive in these unfamiliar surroundings. Countless haciendas have been purchased by passionate conservationists, both foreigners and nationals, and turned into stylish homes. These charming residences blend artisanal craftsmanship with antique objets d'art and modern furnishings. Stunning outdoor spaces, which include sprawling gardens and swimming pools, enliven these outstanding properties. In the towns of Mérida, Izamal, and Valladolid, picturesque colonial houses, surrounded by lush vegetation, have also undergone restoration while retaining their original charm.

Journeying through Yucatán is an extraordinary voyage of sights and sounds that transports you back in time. It's an escape to a world of dreams and grandeur, comparable to delving into the magical realism you'd usually find in a novel. Intended to inspire, this book is about beautiful houses and haciendas, and the one-of-a-kind homes of exceptional people. It has been a joy to create. I hope that readers will relish their journeys through it as much as I did mine.

Susana Ordovás

# COUNTRY

It was a hot, balmy day and the vast Yucatán sky stretched above me in an immaculate cerulean blue. I peered through the cloudy windscreen of my car as it crept along the narrow country road surrounded by brush and scraggly trees. I briefly questioned if I was headed in the right direction. Suddenly, Google Maps redirected me to the left and into a wide, flat field. In the distance, I spotted a derelict neoclassical building obscured by overgrown trees. To the left, on the far side of the field, stood an austere edifice dating back to the period of Spanish rule. Crowned with rows of sagging arches, its weathered patina conveyed the relentless passage of time. I had arrived at Hacienda Uayalceh, and was immediately captivated. Thus began my love affair with Yucatán and its haciendas.

While the exact number is uncertain, during the henequen heyday from 1880 to 1910, it is believed there were over a thousand haciendas in Yucatán dedicated to the production of the fiber. The agave plant was so valuable, in fact, it was known as *oro verde*, or "green gold". The term "sisal" comes from the fact that henequen was largely exported from Sisal, Yucatán's principal port during this period. A handful of families controlled the trade and spent their fortunes on the finest European furniture, china, and crystal, which they used to decorate the haciendas they transformed into a bewildering variety of styles, from medieval to Moorish and Renaissance to Baroque, and sometimes a combination thereof. The result of this concentration of immeasurable wealth can still be seen today in the countless haciendas scattered across the region.

In the 1940s, misfortune engulfed this land of richness. The invention of synthetic fibers led to a drop in global demand for henequen. Additionally, competition from other regions of the world that offered cheaper alternatives, along with political factors such as the Reforma Agraria—land distribution policies imposed by the Mexican government—contributed to the demise of the industry. With no other sources of income, most haciendas were abandoned and left to perish in the unforgiving jungle.

For decades, the landholdings remained in a derelict state—until the early 1990s, when some of them were painstakingly brought back to their former splendor. Despite these efforts, most of Yucatán's stately homes and haciendas still lie abandoned, reclaimed by the indomitable wilderness. They remain a romantic reminder of the difficult yet fascinating history of this remote corner of Mexico.

Yucatán
Peninsula
Gulf of Mexico
Campeche
Campeche
Haciendas

Río Lagartos
National Park
Mérida
Paraíso
Komchén
Cuzumal
San Antonio
Millet
Tekik de
Regil
Plantel Matilde
Sac Chich
Yaxcopoil
Petac
Subin
Xukú
Uayalceh
Izamal
Uxmal
Valladolid
Cancún
Yucatán
Quintana
Roo
Tulum
Biosphere
Reserve
Sian Ka'an
Caribbean Sea

# HACIENDA XUKÚ

## *Seyé*

As the sun hung low in the sky, casting a warm, golden glow, I embarked on a journey to a remote location nestled deep within the lush jungle of the Yucatán Peninsula. I drove down a narrow dirt road that wound its way through the dense vegetation, the tires of my vehicle gently crunching against the gravel and earth. The thick foliage on either side formed a verdant tunnel, occasionally allowing slivers of sunlight to penetrate the path ahead. The trail seemed to stretch endlessly—until, finally, it opened to reveal a clearing, and there it stood: Hacienda Xukú.

I drove through the weathered, arched gateway, and the expansive grounds unfolded before me, displaying a carefully cultivated landscape that seamlessly merged with the untamed wilderness I had just left behind. Vibrantly plumaged peacocks roamed free and a pair of horses grazed, their manes tousled by the gentle breeze. Nearby, a donkey brayed softly. The imposing, crimson-colored structure, constructed with ancient stones from the Maya vestiges that populate the property, exuded a sense of quiet grandeur. Except for the long rows of Gothic-inspired arches that line both the front and rear of the main house, the architecture is marked by simplicity, devoid of ornate embellishments or ostentatious details.

The meaning of the word *xukú* is an enigma. Although elderly people who once lived and worked on the hacienda believe it refers to *xuk'úh*, a Maya word that means "corner of the Moon", there are many theories to its true definition. Like similar landholdings in Yucatán, Xukú began its long life as a cattle and agricultural farm in the seventeenth century. It was transformed into a robust henequen plantation in the nineteenth century when the interest in the agave fiber soared worldwide. Once the demand plummeted, a hundred years later, Xukú faced a bleak fate, and the once prosperous holding found itself grappling with economic decline. Abandoned, it fell into disrepair and its commanding structure faded with neglect.

Twenty-seven years ago, Juan Almazán acquired the forlorn estate, which lay entangled in vegetation and ruin. The intrepid businessman embarked on a mission to restore the dilapidated site to its former splendor, enlisting the expertise of talented architectural designer Emmanuel Picault and a dedicated team of specialists. With determination, Juan likened the meticulous restoration procedure to a piece of hand embroidery, underscoring its intricate artisanal nature. A work in progress, this endeavor serves as a testament to his unwavering commitment and the remarkable resurgence of Hacienda Xukú.

At the heart of this isolated domain stands an ancient Maya pyramid. I carefully ascended the steep steps, and as I reached the summit, I saw a breathtaking panorama spread out before me: the vast sea of green that was the tropical forest extended in every direction. Ensconced in the middle, standing proudly, was Hacienda Xukú.

# HACIENDA PARAÍSO KOMCHÉN

## *Motul*

Upon arriving at Hacienda Paraíso Komchén, I am greeted by rows of magnificent multicolored bougainvillea and towering palm trees. The main house is ensconced in a tropical garden of ceiba and ficus trees, and vibrant heliconias that add a splash of color to the luxuriant landscape. Providing ample shade, the porch of the house is graced by a majestic guanacaste tree, or *pich* in Maya, with cascading greenery that stretches wider than the tree's impressive height. Stepping onto the porch to the notes of a Beethoven piano sonata is a delightful experience, even more so when I realize that the melodic notes floating from within the house are live, being played on a Steinway by my host, Roberto Abraham. Nico, one of three white cockatoos residing on the lush estate, perches gracefully on his shoulder.

Paraíso Komchén, once the former outbuilding of neighboring Hacienda Komchén de Martínez, was purchased in 2009 by Roberto Abraham, who, without any prior intention of owning a property in the countryside, became enchanted with its beauty and energetic splendor after being invited to spend the day there by a friend and the previous owner's sister. The picturesque estate was in excellent condition when he acquired it and, after making a few minor stylistic alterations, he proceeded to fill it with antique furniture and objets d'art he searched far and wide for. In the *sala*, a white alabaster archangel, discovered in an antique shop, stands guard on a marble-topped round table. It shares the space with a dining table, chairs, and two sideboards procured from the daughters of Miguel Peón Casares, the original owner of El Pinar (see page 160). An imposing French armoire and bed, also from the iconic mansion in Mérida, take pride of place in the only bedroom of the main house.

Roberto is the State of Yucatán's main cultural promoter, successfully spearheading and promoting two of its largest cultural projects in recent years: El Gran Museo del Mundo Maya in Mérida (The Great Museum of the Maya World) and Centro Nacional de la Música Mexicana (National Center for Mexican Music). He is also president of the Yucatán Culture Foundation and is part of several collegial bodies that promote and strengthen the arts in the State of Yucatán.

THE
AERMOTOR
CHICAGO

# HACIENDA SUBIN

## *Subinkancab*

Snuggled up against the unruly jungle, under vast skies, a crumbling, pink hacienda sprawls in the tropical sun. One of the oldest of its kind, Hacienda Subin was built in the 1680s as a large cattle farm, and like most haciendas that survived into the nineteenth century, it was converted into a henequen plantation. After the industry plummeted and production in Hacienda Subin came to a grinding halt, the building was eventually converted into a village school, until it was abandoned and inevitably devoured by the indefatigable vegetation.

Laura Kirar and Richard Frazier were in Yucatán for a wedding in 2008 when, while driving aimlessly through the countryside, they stumbled by chance upon some doomed ruins. Intrigued, they climbed over a waist-high wall and fought their way through thorns and vines to discover a spectacular structure of thirteen double-molded Moorish arches and an enfilade of disintegrating rooms. The portico had been swallowed up by creeping plants and trees, the roof had all but disappeared, and full-grown ceiba trees had put down roots in the chapel. A sudden shower graced the sky, followed by a rainbow. Laura turned to Richard and quipped, "It's a good omen. The house is supposed to be ours." Despite not knowing if the neglected estate was even on the market, they toasted each other with lukewarm Coronas, indulging in a bit of wishful thinking.

As luck would have it, the next day they found themselves driving on the same less-traveled road. This time, there was a brand-new BMW in front of the waist-high wall. Intrigued, they pulled up next to the car and waited until a tall, handsome man appeared and, to their astonishment, introduced himself as the owner, and the woman beside him as his real estate broker. "And 48 hours later, we were signing papers that would change our lives forever," reminisces Laura.

Driven by this quasi-divine intervention, the couple set about restoring the forsaken ruin to its former glory. "There were cracks compromising the arches. Very little remained of the original woodwork, save for a few remnants of doors and windows. And yet, the place had an air of *The Secret Garden* to it. It was rough and romantic and full of potential." By working on it arduously over the years, the couple have transformed Hacienda Subin into their prepossessing home. Here, accompanied by their fourteen adopted dogs, they throw spontaneous weekend parties, lounge around the new pool, cook with friends, and play pétanque at the end of the day when things cool off.

HOME
01

# HACIENDA YAXCOPOIL

## *Yaxcopoil*

Magical, extraordinary Hacienda Yaxcopoil. I was lucky enough to see it devoid of people the first time I visited it, and fell hopelessly in love with its echoing rooms and creaking furniture. I spent the entire day roaming its musty interiors and atmospheric grounds, discovering the labyrinthine dwelling, closed to visitors and open to me, with a sense of awe. I walked aimlessly through enfilades of history-laden halls, the soft ringing of my footsteps and the occasional groan of an old door filling the air. Outside, there was the soft whisper of rustling palm trees as they swayed capriciously in the lazy breeze.

Built with ancient stones gathered from the pre-Hispanic ruins found abundantly in the vast grounds, Yaxcopoil, which means "place of the green poplars" in Maya, started its protracted life in the seventeenth century as a humble cattle and corn farm. It slowly grew in size and importance, and in the year 1800, Andalusian architect Santiago Servián was summoned to erect a Moorish influenced double arch in the entrance of the hacienda, as a symbol of its grandeur.

Though Yaxcopoil changed hands various times over its long history, Donaciano García Rejón, whose portrait hangs proudly on the wall of one of the *salas*, next to that of his wife Mónica, purchased the hacienda in 1864, along with its main house, outbuildings, 25,000 acres of land, 2,000 cattle, numerous horses and mules, and 250 acres of henequen. Thereafter, and due to the soaring worldwide demand for henequen, Yaxcopoil was swiftly transformed into a successful henequen plantation, with all other agricultural activities abandoned for the more profitable endeavor of producing the highly coveted agave fiber. At the turn of the century, during a time of immense prosperity, the hacienda was modernized, and its buildings were enlarged and embellished. With no expense spared, all installations were improved and a new workshop with a striking neoclassical façade was erected.

However, this time of bonanza was short-lived. Social changes brought about by the Mexican Revolution, as well as the introduction of synthetic fibers into the world market, caused a rapid decline in the demand for henequen. Exactly a hundred years after the industry exploded in the 1880s, Yaxcopoil shut down its henequen production for good. The sprawling property is now owned by a direct descendant of Donaciano García Rejón, Miguel Faller, who entered wholeheartedly into its restoration when he became its sole proprietor in 1986. In his constant, arduous labor of love, carried out in order to preserve things as they were, Faller reminisces about the past. "I used to come at weekends as a child with my cousins and we'd roam the countryside. I have great memories of swimming in the cenotes," he says with a trace of nostalgia in his voice. Hacienda Yaxcopoil, gracious and stately, and now open to the public, is a poignant remnant of an earlier, more industrious era.

DE MEDICINA Y CIRUGIA
ATLAS 2ª PARTE
TOMO 2
TOMO 3
HISTORIA NATURAL
LA VUELTA AL MUNDO
1862
DICCIONARIO DOMESTICO
HISTORIA UNIVERSAL
SAGRADA BIBL
DICCIONARIO ENCICLOPÉDICO DE AGRICULTURA
TOMO III
TOMO IV
TOMO I
TOMO II
TOMO VII
TOMO V
TOMO VI
SUMA TEOLOGICA
E. RECLUS NUEVA GEOGRAFIA UNIVERSAL
ASIA ORIENTAL
CHINA COREA JAPÓN
AMÉRICA CENTRAL
LA TIERRA
WHITEHOUSE COOK BOOK

# PLANTEL MATILDE

## *San Antonio Sac Chich*

Colossal, imposing, and austere, Javier Marín's home in Yucatán is indeed otherworldly. Rising above the canopy of a ragged jungle, the concrete immensity sits omnipotent in a vast clearing, dominating the landscape with its symmetrical, clean lines and expansive proportions. Like the ancient pyramids built by the great Maya civilization, in which nothing is superfluous, Plantel Matilde carries the strength of simplicity and the power to convey almightiness.

On a late afternoon in February, I made my way through the tiny village of San Antonio Sac Chich and down a bumpy dirt road enveloped by a thick curtain of dzidzilche, jabín, and ceiba trees. After a short while, I came upon an opening, and before me stood Plantel Matilde, a monumental building of such constrained elegance that I was stopped in my tracks, struck with wonder. The surroundings seemed to embrace silence; only the faint rustle of leaves caressed by a gentle breeze could be heard in the distance.

Javier Marín, artist and sculptor, greeted me warmly with his usual discreet demeanor, his humble nature in contrast to his dramatic, larger-than-life head sculptures. Built in 2010 in collaboration with his brother, architect Arcadio Marín, Plantel Matilde is his refuge from the world, a sanctuary where he finds solace and connects with nature. "Just like a blank canvas, this is a space for contemplation and meditation, where I get to spend time with myself," says Javier. "It nurtures my spiritual reflections and allows me to imagine and create."

A perfect square, the contemporary edifice surrounds a man-made lake with a small island at its center where vegetation grows, purposely left untouched. The structure, raised slightly above ground level, provides a cavernous area below for the living quarters, which are decorated sparsely with furniture designed by Arcadio and made by local artisans. One's attention is constantly drawn to the objects and artifacts designed by Javier that occupy each surface, thoughtfully grouped but never cluttered. The bare walls suggest a level of simplicity belying the careful thought that has gone into each space.

As we sat outside chatting idly, drinks in hand, the heaving sun began to set, casting dramatic shadows around us. I looked around, admiring the isolated, spiritual-like oasis, and I understood the sense of peace Javier was so eager to capture when he first imagined the place.

# HACIENDA SAN ANTONIO SAC CHICH

*San Antonio Sac Chich*

"I was on a very personal quest for a better understanding of color and light. In Yucatán I found color, used uninhibitedly, freely," says Jorge Marín wistfully, as we stand in the shade of one of his monumental winged sculptures. In the background, his splendid hacienda sits amidst the unruly garden he so lovingly tends. "The scent of this region also captivated me, blending with the colors, reminding me of one of Paul Gauguin's artistic stages," he admits as he bends down to caress the muzzle of one of his faithful dogs.

Jorge Marín is a figurative artist best known for his depictions of bronze mythical creatures, including centaurs and masked angels. Born in Uruapan and based in Mexico City, where his studio is located, Jorge spends as much of his free time as possible at his hacienda in Yucatán. Here, he plays an active role in the community through his non-profit organization, the Jorge Marín Foundation, which aims to promote and support social causes that are close to his heart, such as education, migration, and the reclaiming of public space.

Located in the village of San Antonio Sac Chich, this hacienda was built in the mid-1800s, during the birth of the "green gold" henequen era, when the agave plant native to the peninsula transformed the region's society and economy. The rich interiors, presided over by imposing rows of Corinthian columns that were erected when the hacienda was constructed, are a clear reflection of his aesthetic and point of view. A free spirit who expresses himself daily through the art he creates, the soft-spoken artist has found a way to also do this through the spaces he decorates. Each corner, each tiny nook in his home is a faithful reflection of his wonderfully eclectic taste. "Every person's story can be told through the objects they hold dear. For me, the individual value of each piece is secondary. What matters is the collective narrative they weave and the atmosphere they create as a whole."

As we wander around the grounds, followed by his loyal tribe of adopted dogs, he explains, "I would love for everyone to have the opportunity to experience creating or tending to a garden. It's akin to sculpting or painting, but without the constraints of time. It fosters a deeply enriching connection with nature, allowing us to discover our place in this magnificent world."

MARIA
MADRE
DIOS

# HACIENDA CUZUMAL

## *Samahil*

It was near the end of the rainy season and, as I wilted under the stifling heat, I took in the wonders of the charming hacienda hidden in the oblivion of a thick jungle. The former tumble of stones, once buried under a tangle of hungry vines, now shone proudly with new magnificence.

The history of Hacienda Cuzumal is eerily similar to that of other haciendas in Yucatán. After experiencing years of prosperity, thanks to the wave of wealth that swept over the region during the "green gold" era, or henequen heyday, it was unmercifully abandoned once the industry collapsed in the mid-twentieth century. Discarded and left to decay, the noble structure was overtaken by the merciless environment, and devolved into a mound of stones and shapeless rubble.

Unlike other, less fortunate properties engulfed by Yucatán's tropical rainforest, Hacienda Cuzumal found a new lease of life in 2003, when it was purchased by the Hernández family and put in the care of a team of expert architects whose mission was to rescue the old building and return it to its former glory. Composed of stones belonging to a nearby pyramid built centuries ago by the skillful hands of Maya masons, the original structure had almost entirely disintegrated. Believed to be inspired by the Governor's Palace in Uxmal, the main house was rebuilt and the surviving zapote wood beams (known as *sak yá* in Maya) that had stood the test of time were carefully restored and put back in place. To the delight of everyone involved in the rehabilitation process, once the undergrowth was cleared away, delicate stenciling appeared on the peeling walls. In what was once a sprawling patio located behind the main building, a large pond with water lilies was installed, and around it, three elevated villas were erected to house comfortable suites for guests.

In what was no small feat, due to its isolated location in the impenetrable wilderness, the hacienda was painstakingly brought back to life over the course of three long years. Now serving as a tranquil retreat under the management of Private Haciendas—who also oversee neighboring haciendas Tamchen and Tixnuc—this historic sanctuary offers a rare opportunity to completely disconnect from the outside world and immerse oneself in the embrace of nature.

# HACIENDA PETAC

## *Petac*

When Dorothea, fondly known as Dev, and Chuck Stern arrived in Yucatán in January of 2000 and saw it was full of abandoned haciendas to choose from, they knew they wanted one for themselves. Having crawled all over northern Mexico, it took them barely a week to decide that Yucatán was where they wanted to fulfill their dream of having a holiday home. They drove around for days looking for the right hacienda until they came upon a crumbling ruin built on the vestiges of a Maya site in the tiny village of Petac. The beauty of the location and its string of Moorish-style arches proved to be irresistible. However, not everyone understood their vision. As the couple from Texas explained to me: "No one who saw the pictures could believe we wanted to take on such a huge project. Our son thought we were nuts. Our family thought we were nuts. Our friends thought we were nuts. Apparently, only we could see its charm and potential."

After purchasing the forlorn eighteenth-century property—which had seen better days, first as a cattle farm and then as a henequen plantation—they threw themselves into its rehabilitation. "The first year was a disaster, because we didn't know what we were doing," admitted a bemused Chuck. However, things took a turn for the better after they acquired the expert help of architectural duo Salvador Reyes Ríos and Josefina Larraín, and over the course of the next two years, they worked tirelessly to bring the place back to life. All the original bones were preserved, with the façades restored, many doors and beams rescued, and original colors reinstated where possible. Although only a very small patch of floor tile was salvageable, they were able to have the original tile design made into a new mold and manufactured locally. The kitchen, once an open, wood-burning hearth, was transformed: "We reimagined it as a more welcoming space. The table was purchased in Puebla and the tiles were bought in Mérida." Here, traditional specialties such as *panuchos*, *frijol con puerco*, *poc chuc*, and *cochinita pibil* are now whipped up by women from the village who help take care of the hacienda.

As they neared the completion of the works, a Category 5 hurricane swept in, threatening to destroy the new gardens together with thousands of dollars of investment. Chuck vividly described the experience: "As the hurricane approached the hacienda, it sounded as if dinosaurs were stomping towards it as trees fell. When it passed, our first view was disturbing, but also rather miraculous." The structure of the buildings had remained intact and two trees had fallen into each other directly over an arch, preserving it. "Our gardeners looked at what could have been and declared that the *Aluxes* had protected us," said Chuck. *Aluxes* are Maya mythological creatures, similar to goblins or elves, that watch over the fields and the land. "The gardeners immediately started righting trees and replanting. Almost everything was saved."

# HACIENDA UAYALCEH

## *Abalá*

Two lean horses stood motionless, except for the occasional swish of their tails. The relentless heat seemed to subdue all movements: no man, creature, or leaf stirred; the only sign of life was the faint sound of *jarana* music coming from a distant transistor radio. I walked up the crumbling steps to the frail, wooden front door of the old building and knocked. Once, twice, three times. No response. Disheartened, I stepped back to take in the captivating sight of the decrepit structure.

After a minute or so, the sound on the radio stopped and I heard footsteps approaching the door. A short, middle-aged man with a stocky build opened it and smiled apologetically. "*Bienvenida a Hacienda Ualayceh*," he said, gesturing me in. I was soon to learn that his name was Fernando Cauich Cetz, known affectionately as Nando, and he was the caretaker of the semi-abandoned, decaying estate where his family had been laborers for generations. He himself had worked in the fields and at the hacienda from a young age, when it was still a working plantation.

*Uayalceh*, which means "land where the deer abound" in Maya, has belonged to the Peón family since the eighteenth century. It was first purchased by Alonso Manuel Peón Valdez, a high-ranking military officer and politician from Asturias, Spain, who served as acting governor of Yucatán on three occasions. For the first hundred years of its existence, the landholding operated as a cattle and agricultural farm. However, in the mid-nineteenth century, it transitioned to exclusively cultivating and processing henequen.

Frequently referred to as "green gold", henequen brought enormous wealth to the Peón family, who owned multiple *estancias*, or large estates, and participated in all aspects of the trade. At the turn of the nineteenth century, the Peóns played a pivotal role in fostering Yucatán's development. By adeptly repurposing their holdings into henequen-producing enterprises, they not only contributed significantly to the local economy but also were key in facilitating Yucatán's global engagement and outreach.

After the demand for henequen dwindled in the mid-twentieth century, activity diminished over the years, until it was suspended for good. After a life of staggering opulence, Alonso Luis Peón Bolio sold the forsaken property in 1982 to his son, Alonso Peón Martínez, who would visit the place every Thursday. The nostalgic proprietor would wander aimlessly through the echoing rooms and spend hours in his rocking chair on the outdoor gallery, observing the untamed landscape through a string of delicate columns and arches.

Alonso, who never married or had children, passed away on September 15, 2018, leaving the estate to the remaining twenty members of the Peón family. After being shown around by Nando that day, I found Uayalceh, despite its years of neglect, to be deeply moving—and still beautiful in its own way.

# HACIENDA SAN ANTONIO MILLET

## *Tixkokob*

Many *hacendados* of New Spain had their family coat of arms depicted on the front of the main house of their haciendas, as proof of their grand lineage or that of their ancestors. Such is the case of Hacienda San Antonio Millet, purchased in April 1881 by Alvaro Peón de Regil and his wife Candelaria Peón Castellanos, Countess of Miraflores. As the only title ever to be granted in the Yucatán Peninsula—by the Spanish Crown, in 1689—the Countess of Miraflores was proud of her aristocratic heritage, and had her family coat of arms placed on the new turret of the *casa principal* and above the door of the newly commissioned chapel.

During this time of economic bonanza due to the thriving henequen industry, the hacienda underwent a gradual but important refurbishment—one worthy of royalty. With no expense spared, construction materials were imported from France and the two main buildings were embellished with striking turrets and battlements to emulate medieval and Renaissance structures.

It was a late November afternoon when I first drove through the gates of the imposing property. Two enormous carob trees cast dramatic shadows over the manicured lawn and onto the sweeping steps leading up to the main house. There to receive me was Eduardo Calderón, whose love for Yucatán prompted him to buy the sprawling estate in 1996. His wife Cándida Fernández, director of Fomento Cultural Citibanamex, had been doing research for a book on Mexican haciendas when they came across the impressive seventeenth-century estate in the municipality of Tixkokob. The casco, or the main house and outbuildings, had long been abandoned and were in a state of semi-ruin, and the grounds were scorched and scarred, with nothing but a few sad trees, wistful survivors of grander days.

Eduardo took it upon himself to restore San Antonio Millet to its former glory. The first major step in the process, which was to take thirteen long years, was the reconstruction of the caretaker's house. The vast gardens were then brought back to life. The machine room—once a crucial hub that housed steam-powered machinery such as a decorticator, used to extract fiber from agave leaves—had its structure strengthened to prevent its façade from collapsing. Soon after, expert restoration work was carried out on the intricate stenciling on the walls of the main house. As a result, San Antonio Millet is one of the only haciendas in Yucatán that retains its original stenciling. Eduardo and his team focused their final efforts on the chapel—rehabilitating the altar, the confessional, and the railings, and placing above the pulpit an abat-voix they rescued from an antique shop in Mérida.

Eduardo, Cándida, and Eduardo's brother, José, who joined halfway through the restoration, are now proud owners of one of Yucatán's most remarkable landholdings. Their intention was to have a haven, a place to get away from everything. Instead, it has become a mission, a way of life.

# HACIENDA TEKIK DE REGIL

## *Tekik de Regil*

Tekik de Regil is one of those haciendas you find only in Yucatán, where properties were designed with a bewildering eclecticism fueled by the fantasies of their owners—a phenomenon I witnessed firsthand when I visited the grand estate one damp day during rainy season. As I approached the *casa principal*, or main house, I was captivated by the colorful frescoes that adorned the exterior walls of the expansive Palladian building, transporting me back to the ancient world of the Maya. Painted by local contemporary artist Carlos Millet, also known as "Calocho", who drew inspiration from Villa Farnesina in Rome, the murals skillfully intertwine remnants of a distant past with the vibrant essence of the Yucatecan wilderness. Tropical fruits and exotic wildlife come to life on the walls, creating a tableau of colors and life. Vibrant parrots and other animated birds, with feathers of scarlet, azure, and gold, seem to perch on door frames. Among the foliage depicted in the frescoes, playful monkeys leap and swing, adding a touch of lively movement.

Believed to have been founded towards the end of the sixteenth century, Tekik de Regil stands as one of Yucatán's oldest landholdings. Named after the Maya word *kik*, which means "place of blood", it started off as a cattle farm, but was converted into a plantation as demand for henequen surged worldwide. It was during this pivotal time that the owner, Pedro Regil Casares, increased the size of the property and modernized production by investing in cutting-edge machinery. He also sought the expertise of Italian architect Alfonso Cardone to rebuild and transform the prosperous hacienda into a lavish estate.

Isolated by its own magnitude, Tekik de Regil created a self-sufficient community within itself. Safeguarded by tall walls, the *casco* was the heart of the hacienda. It consisted of the *casa principal* and the chapel, workshops, machine room, stables, workers' quarters, and *tienda de raya*, or general store, where laborers, who received their wages in the form of vouchers or tokens, could exchange them for goods. The importance of the productive aspect of the hacienda was evident in the closeness between the *casa principal* and the outbuildings.

When the henequen industry experienced a sharp decline in the first half of the twentieth century, Tekik de Regil faced financial challenges, marking the end of its golden era. The hacienda was abandoned, falling into a state of disrepair. A major monument to a fascinating time, the decaying estate proved to be indestructible. Purchased by the Hernández family in 2002, it was restored with care over the course of three long years.

As I stepped into the empty halls, I felt the air heavy with humidity and the scent of aged wood. A soft light filtered through the windows and doorways, casting hues on the weathered *pasta* tiles. The walls, adorned with faded stenciling, bore the marks of time, and I found myself transported to a glorious era in which life once thrived.

# TOWN

A few years ago, the thought of writing a book on Yucatán would not have crossed my mind. While Mérida is a charming and historical city, it lacks the Baroque designs of Puebla and is a far cry from the vivid metropolis of Mexico City, with its world-class museums and diverse food scene. And compared to the rich artisanal heritage of Oaxaca City or the busy markets of San Cristóbal de Las Casas, Mérida might, at first, appear understated. But, as they say, first impressions can be deceiving. Unlike other places in Mexico that flaunt their beauties to the world, Mérida's true treasures lie concealed behind tall walls, hidden from prying eyes. I was to make this discovery when I traveled extensively through Yucatán in 2020 and 2021. To my surprise, I encountered an array of one-of-a-kind homes in the towns of Mérida, Izamal, and Valladolid. Yucatán had flourished during my absence of twenty-odd years, and I found myself captivated.

Although the houses shared on these pages have the same basic characteristics of many urban abodes in Yucatán, most of the featured homes are owned by individuals in the creative fields and were designed and executed by the owners themselves, without the involvement of third parties. These proprietors don't adhere to a specific design style; instead, each building reflects the tapestry of its owners' diverse lives and cultural experiences. Whether humble or grand, each home embodies a personal vision rather than conforming to trends, rendering each hacienda unique and unforgettable.

But the true allure of these places lies in the fact that everyone who has chosen to make a home here has brought their own fantasy with them. Today's homeowners have reimagined houses that once resonated with the laughter of long-gone Yucatecans. They have revived ruined *casonas*—or any abode that captured their imagination and fit their budget—in the dusty *centro histórico* of Mérida, as well as unassuming *casitas* and sprawling *mesones* in the charming towns of Izamal and Valladolid, and created their own corner of paradise. Each of these homes has been transformed by the magic of Yucatán, so much so that they could, indeed, exist nowhere else but here.

IZAMAL
Kinich Kak Moo
De los Remedios Chapel
CASA DE MADERA
Citilcum Church
CASA DE LOS ARTISTAS
Plaza Izamal
Itzamatul
CASA AZUL
CASA DE LOS SANTOS
San Jerónimo Church
MESÓN DE MALLEVILLE (VALLADOLID)
Santa Clara de Asís Kimbila Church
Ex convent of San Antonio de Padua
Habuk Izamal
De los Indios Chapel

# MÉRIDA

QUINTA LOS ALMENDROS
EL PINAR
Paseo de Montejo
Tercera Orden Church
CASA MARGARITA
Government's Palace
CASA SERRANO WILLSON
Santa Lucía Park
San Ildefonso Cathedral
Municipal Palace
CASONA DE LA MEJORADA
CASA 65
Casa de Montejo
Candelaria Church
CASA VERSAILLES
CASA BROWN
The Arch of San Juan
XUNA'AN KAB
CASA SAN CRISTÓBAL
San Juan Bautista Church
CASA ERMITA DE SANTA ISABEL

# QUINTA LOS ALMENDROS

## *Mérida*

Magnificent homes, built during the heyday of the henequen boom, abound in Mérida. Elaborate façades hint at the sumptuous interiors and the riots of color and ornaments that adorn these stately dwellings. However, the candidus exterior of Quinta Los Almendros, covered in a pristine layer of white paint, offers no hint of the extravaganza of robin's-egg blue concealed within its splendid walls.

The beguiling mansion, hidden from prying eyes behind tall gates, was built for the Ferráez family by prominent local architect Manuel Amábilis Domínguez, shortly after his return in 1912 from studying in Paris. This exciting project allowed the budding designer to put into practice what he had learned at the École Spéciale d'Architecture, mixing disparate architectural styles such as beaux arts, Caribbean, and antebellum. At once theatrical and distinguished, playful and grand, Quinta Los Almendros was configured as a series of interconnecting indoor and outdoor courtyards, allowing currents of air to freely circulate and alleviate the oppressive heat.

Years later, Los Almendros, which translates to "the almonds", was converted into a school and attended by none other than Manuel Barbachano, the great-grandson of the famed five-time governor of Yucatán, Miguel Barbachano y Tarrazo. Destiny had it that, having fallen in love with the building as a student, Manuel, who later became known for his illustrious career in the Mexican film industry, was able to purchase his former school for his Mexico City–based family in the 1960s.

On a cloudy day in February, I was received by Manuel's daughter Isabel, her husband Federico, and their daughter, who shares the name Isabel. I meandered through the stately six-bedroom stuccoed mansion; it came as no surprise to discover that the interiors and much of the furniture had been masterfully designed by Arturo Pani, commissioned by Manuel and his wife Teresa to restore the estate.

Considered the star decorator for the Mexican elite of the mid-twentieth century, Pani won over an upper-class clientele who appreciated the vivacity and exuberance he brought to interiors. Throughout every room, one could see his expertise in using innovative materials to reinterpret traditional styles, offering an idiosyncratic yet sumptuous take on historical forms. Intermingling tastefully with Pani's pieces were exquisite antiques, such as an eighteenth-century console, a ceramic Venetian fountain spurting water into a pond, and a gilded Eye of Providence in the master bedroom.

The historical property continues to be filled with family, friends, meals, and laughter. Since their father Manuel passed away in 1994, Isabel, her twin sister Teresa, and their brother Francisco have diligently preserved Quinta Los Almendros. With their unwavering dedication, the mansion remains intact and provides the perfect setting for gathering in their beloved Yucatán.

# EL PINAR

## *Mérida*

Like European aristocrats of old, José Trinidad Molina had a regal vision of the life he wished to lead. A castle of some sort was key. So, in 1987, the Yucatecan entrepreneur of modest origins—who made his fortune in the tourism and shipping industries—acquired El Pinar and turned it into his lavish abode.

A striking, chateau-like mansion inspired by sixteenth and seventeenth-century French Renaissance architecture, the dwelling was erected by its original owner, Miguel Peón Casares, in 1904, when wealthy *hacendados* built their lavish homes in a time of economic prosperity brought by the heyday of the henequen industry. No expense was spared in the building of the three-story residence, with Carrara marble imported from Italy, slate tile for the roof shipped from France, and tin cornices for the attic windows purchased in Philadelphia. Even basic construction materials, such as the red bricks used for the walls, were brought from Europe. The Concha del Peregrino, the universal symbol for the Camino de Santiago, is said to have been the inspiration behind the scallop shells depicted on the delicate pink façade and the elaborate stucco ceiling in the dining room.

When the Molina family purchased El Pinar, it was melancholy and worn, but had great promise. Undaunted by such an endeavor, José Molina and his wife, María Lucía, lovingly restored it to its former glory. Silk moiré and damask curtains were hung from windows, fine handmade quilts were spread over beds, and the house was hung with Murano chandeliers, nineteenth-century Sèvres porcelain, an antique Jacobean Revival dining table and chairs, and a Blüthner piano, all pieces the couple had acquired over their decades-long marriage. The devoutly Catholic pair, who had raised twelve children, filled corners and covered walls with religious artifacts and family portraits.

Both have since passed away, yet everything remains in place and intact, tended carefully by their descendants. El Pinar, named after a row of cypress trees that once wrapped around the property, stands proudly to this day as a symbol of a splendid bygone era.

# CASA SERRANO WILLSON

## *Mérida*

On a vibrant autumn evening, I walked through the *barrio* of Santiago, in Mérida's *centro histórico*, to the residence of David Serrano and Robert Willson. Its minimal neoclassical front belied the dramatic spaces I was soon to discover within its solid walls.

In exploring the restored colonial home and admiring its myriad of treasures thoughtfully displayed, the biggest surprise was the revelation that David and Robert moved here in recent years. The house, with its specially commissioned, site-specific pieces of furniture and carefully curated art and objets, suggests a lifetime of layering that can't be pulled together in a matter of months. Yet that is just what happened here, made possible by the couple's discerning taste and impeccable eye for detail.

Born in the dusty desert border town of Mexicali, in Baja California, David's journey took him to Los Angeles, where he began painting professionally, becoming an integral part of the city's Chicano movement. In contrast, Robert forged a successful path in the culinary business in Los Angeles. After a few decades, he decided to leave that behind to fully dedicate himself to managing Downtown, a prosperous twentieth-century design gallery that he had cofounded with David years earlier.

Despite their fruitful careers, David and Robert yearned for a simpler lifestyle. In 2011, after planning an exit strategy, they acquired the house of their dreams in Mérida. The allure of tall beamed ceilings, a monumental stone wall at the rear of the property, and the three flamboyant trees in the grounds had drawn them irresistibly to this place. As the ambitious restoration of the centuries-old dwelling took shape, Robert supervised the process diligently, traveling to Mexico every six weeks. David, on the other hand, chose to keep his distance until the project's completion, focusing on the furniture, artwork, and objets d'art that would later adorn their new home. It wasn't until the containers arrived for installation that he finally made an appearance in Yucatán to witness the realization of their shared vision.

As I progressed from room to room, I was impressed by variety of the displays. From a pair of bleached-oak 1940s club chairs sourced from the south of France to a red oversized lamp designed by Billy Haines and a Venetian grotto chair acquired in New Orleans, every item had been thoughtfully curated. Curious, I inquired about the pieces that held a special place in their hearts. They expressed their fondness for the large, colorful painting in the first *sala*—a commission from local artist Irvim Victoria—as well as their large collection of books, and an unusual marble *pietra dura* table, which serves as both a cabinet of curiosities and a display for *nature morte* objects. I could not help but be moved by the wonderful juxtaposition of items they had collected over time. It was like a visual poem, resonating deeply with the eye.

MODERN
LIVING IN STYLE PARIS
VILLE VENETE
OBJECTS:USA
POTSDAM
VANITY FAIR PORTRAITS
EMIL KAZAZ
SCULPTURE
LUXURY HOUSES
DRESSING THE HOME
ROMAN GARDENS
TONY DUQUETTE
MORE IS MORE
TONY DUQUETTE
PUTMAN STYLE
MODERN INTERIORS
TANGER
LOS ANGELES

# CASA BROWN

## *Mérida*

"To us, the children, this house is the epitome of everything Alexandra and James Brown represented," Dagmar tells me as we lounge on two wrought-iron canopy beds amidst a sea of velvet cushions, in the front *sala* of her late parents' home. "Deeply intriguing, mysterious and captivating ... Every object tells a story, it has its purpose and reason to be in this house. And one, like these objects, should feel great pride to be able to observe and be part of the magnificent world that Alexandra and James created confined within these thick, stone walls." Behind her, hanging low on the peeling wall, is a portrait of herself as a little girl, painted by her mother's sister, Julia Condon.

James Brown, the American painter, sculptor, and ceramist, and his wife Alexandra settled in Mérida after living in the countryside in Oaxaca, where they had raised their three children, Degenhart, Cosmas And Damián, and Dagmar. "Why Mérida?" I ask Dagmar as we sip *agua de limón*, or lemon water. "They fell in love with the strangeness of the Yucatán Peninsula," she answers.

Together, James and Alexandra took on the challenge of designing and decorating the old *casona*, which was purchased in 2004 in a state of severe neglect. They built a pool, and added three new sections to the original structure, including a master bathroom in the Hoffmannesque style that features locally crafted tiles with colors inspired by Dzadz, the cenote located on their land just outside the town of Valladolid. Every stone, beam, window, door, column, and tile was either repurposed from trash that James found in dumps, bought for almost nothing from local antique dealers, or made to order by local artisans.

The colonial dwelling displays a captivating blend of colors, thoughtfully curated by James. "The pink of the walls of the outdoor *salón* is a replica from a Venetian palace, from which my father peeled a strip and kept the paint chip in a drawer for years," explains Dagmar. "This drawer still exists, filled with dozens of envelopes labeled with the date and place from which the paint chip was taken."

Dagmar, entrusted with her parents' estate after their passing in February 2020, is quick to point out that life at Casa Brown is never a dull affair, with an unending array of fresh challenges every week – be it rainwater seeping through the roof, tree roots entwining with plumbing, cherished artwork and books succumbing to mold, or unfortunate iguanas falling into the pool and meeting a watery fate. The rainy season in Yucatán also presents its own predicaments. "During the summer months, we board up the doors and windows and pray to Chaac, the Maya god of rain, that he be merciful enough that we may find the house still standing upon our return," muses Dagmar. "I know every person who owns land in the Yucatán deals with the same problems, and with this comes a sense of comfort and acceptance."

MILLER'S
DE KOONING

REMBRANDT
VERMEER
FRANS HALS
ZURBARÁN
MATISSE

# CASA ERMITA DE SANTA ISABEL

## *Mérida*

One might mistake John Prentice Powell's dwelling in the old *barrio* of San Sebastián for just another of the beautiful yet humble Spanish colonial houses along the Camino Real, a road that Empress Charlotte of Mexico herself traveled during her official visit to the Yucatán Peninsula in 1865. And beautiful it is, but it is also so much more. The unassuming, chocolate-colored façade nestled among a long row of similar one-story houses lining the brick-clad street is deceptive, for it hides a singular and magical world.

I cannot remember who introduced me to John. I do, however, vividly recall meeting him for the first time. On a sultry day in Mérida, as I stood outside my hotel under a merciless sun, a dilapidated second-hand Volvo drove up with John behind the wheel. As he flung open the passenger car door and waved me in, I was instantly drawn to his tranquil demeanor and friendly smile. It was the first of many encounters and we soon became fast friends.

Having enjoyed a multifarious career in New York and Paris, John eventually moved to Yucatán after closing the doors to his antique and furniture emporium in New York and purchasing Casa Ermita de Santa Isabel. He arrived in Mérida in 2001 with the first wave of expat creatives, and began to breathe life into the sleepy *centro histórico* by renovating crumbling colonial-era houses. He also cofounded Urbano Rentals, a holiday-rental agency specializing in heritage properties. He named his one-bedroomed casita Casa Ermita de Santa Isabel, after the eighteenth-century hermitage down the road. The charming neighborhood, one of the oldest in town, is a stone's throw away from Arco de San Juan, one of three surviving defensive gates from the seventeenth century.

As we crossed the threshold into John's house, we were welcomed by a soft, soothing light, a forgiving contrast to the blinding sun outside. In what is typical of old colonial houses in Mérida, one enters the front *sala* directly from the street. In the center of the room was a mammoth bakery table John found on the side of the road, leaning against a house, which took eight men to move. Arranged on top was an old Sybilla rug—unfaded over time—and a dusty collection of green-glazed pottery from Michoacán.

As I wandered through the house and into the spacious bedroom, admiring the vermilion wooden beams lining the tall ceiling, I heard the horse and buggies going from their nearby stables to the *zócalo*, or main square. Time stood still and I felt transported to another era.

# CASONA DE LA MEJORADA

## *Mérida*

I remember the first time I stayed at Casona de La Mejorada. After a light dinner, we sat on the patio under the arches counting the stars and talking late into the night. Outside, life raced past, but within the heavy walls of this majestic dwelling, time appeared to stand still, the silence broken only by the movement of leaves coming from the garden, or the distant crack of a firework illuminating the dark sky.

At night, I would lay in bed looking up at the wooden beams high above, a rickety fan purring away, in hopes that the distant rumble of an incoming storm would cool the stifling air. I became accustomed to sleeping with the windows open, only a thin mosquito net between me and the treacherous insects beyond. I'd awake in the early hours of the morning, sluggish from the dreams of a heavy sleep, and hear in the distance the growl of a bus as it went by. On weekdays, the muffled squeals of children playing in the grounds of a neighboring school would drift over the stone wall. Their young voices chanting the national anthem before they'd take to their daily lessons became a constant in my morning routine. It was a hymn I had inevitably memorized by the time my days in Yucatán were over.

To visit this home full of shadows and echoes in the historical *barrio* of La Mejorada is to enter a world of enchantment. The string of connecting rooms, with their original *pasta* tiles, lends the rambling casona an air of timelessness. Hints of color appear behind heavy wooden doors that reveal an enticing trail of wonders, with vintage Guatemalan and African textiles, and Chinese antiques blended into a fascinating, unexpected, and stylish amalgamation. Built in the beginning of the eighteenth century, this typical viceroyal house in Mérida's *centro histórico* is situated next to Parque de La Mejorada. With its solid and grandly proportioned structure, it is only a few steps west of the former Franciscan convent, Parroquia de Nuestra Señora del Carmen, where the tolling of church bells, calling to prayer, resonates throughout the day.

YUCATÁN
MAYA

# MESÓN DE MALLEVILLE

*Valladolid*

Fueled by insatiable curiosity and instinctive creativity, Nicolás Malleville lives a nomadic existence in the constant search of beauty. With an inimitable aesthetic grounded in a sense of timeless luxury, Nicolás travels the world gaining inspiration from imperfection, and collecting ideas that permeate every property he creates. He is also a multitasker: as a hotelier, landscape designer, and perfumer, the founder of the hospitality brand and fragrance line Coqui Coqui has created a lifestyle, alongside his Italian-born partner Francesca Bonato and three children, in which boundaries between reality and fantasy are elusive. Their life in Yucatán and now Bora Bora is truly the alchemy of a wanderlust existence.

Before taking flight to Bora Bora, Nicolás and Francesca lived in the tranquil town of Valladolid, where they purchased and renovated a historic *mesón* from the 1600s. Set elegantly back on a far corner of Parque Sisal—directly across from the former Convento de San Bernardino de Siena—Mesón de Malleville became their home for ten precious years. Offering just a single bathroom at the back of the garden and lacking a central courtyard, which had been lost at some point over the years, the centuries-old dwelling was in need of rescue when the couple acquired it. They respectfully restored it, extending it towards the back and laying their own style over the existing blueprint. They then filled it with antique treasures from around Mexico and pieces Francesca inherited from her grandfather's apartment in Milan.

In the main *sala*, as you walk in from the street, one cannot but notice the imposing presence of a very large Gothic Revival wardrobe, found by Nicolás in a carpenter's workshop, where it had been sitting in a state of abandonment for sixty years. The antique brass beds were bought in Puebla, the clawfoot bathtub was found in an old *casona* in Valladolid, and two nineteenth-century wooden doors were rescued from an old theatre in Mérida. Everything is imbued with story and soul. "For me, every single piece is like a collective archive of my past, of my present and of my future. I don't ever just design things. I fuse ideas that together tell bigger stories. I make my own fantasy," says Nicolás.

Fascinated by travel and adventure, and now living halfway around the world, Nicolás and Francesca remain captivated by the charm of Yucatán. It is here, to this small town between Cancún and Mérida, that they continue to return, and where Mesón de Malleville has become their beloved home away from home.

MODERN ARCHITECTURE

# CASA MARGARITA

## *Mérida*

On one of the centuries-old streets of Mérida's *centro histórico*, just off the historical Paseo Montejo, sits the house of Jenne Maag. Its discreet, faded gray façade, adorned with three nodding palms sitting in large pot-bellied jardinières, belies the Old Hollywood–style interiors that await inside.

The cool, mint-green hues throughout are a welcoming embrace for weary visitors and, not unlike the mansion in the 1950s movie *Sunset Boulevard*, the interior appears as though nothing has changed in decades. Cluttered with memories of former glories, and boasting lofty ceilings, ornate furniture rescued from old opera sets, dramatic candelabras, delicate Murano champagne glasses, and vintage Fortuny fabrics, the grand *casona* appears to be a time capsule resembling the elaborate sets of silent films.

The richness of Jenne's life is reflected in her rooms. An Empire bed from the estate of legendary antiques dealer and designer Madeleine Castaing stands in pride of place in one of the parlors, presided over by a Florentine mirror and a Rose Tarlow table. Taking a rickety elevator to the second floor, one arrives at an exotic, saffron-colored room with angry dragons on pedestals and a sprawling opium bed. These are spaces where the likes of Lawrence of Arabia, Cleopatra and Ivan the Terrible would feel right at home. Passionate about antiques, Jenne has furnished her beloved home with pieces collected throughout her lifetime, most of them shipped from her former villa in Chianti, Italy. Regardless of value and provenance, these are objects that have passed the test of her discerning eye and are well loved.

Originally from Texas, Jenne made her mark on the fashion industry designing highly coveted sportswear for women during the nineties. She has called Mérida her home since 2004, when she purchased the sumptuous house online—and on a whim, without ever seeing it in person. She admits to quickly falling in love with images of the generous atrium, the rows of arches, and the ample terrace with its 39-foot (12-meter) columns. "My friends thought I was crazy, but it was so unusual, so beautiful," she says, without a hint of regret in her tone.

# CASA 65

## *Mérida*

Cuban-born artist Jorge Pardo is renowned for his bold creations. Within his Mérida home, he has employed a palette of vibrant colors and sinuous forms, utilizing both natural and industrial materials. By doing so, he has blurred the lines between art, architecture, and design, effectively demonstrating that a living space can be all these things at once.

When Jorge purchased his walled oasis in the *barrio* of Santiago, in 2013, it was little more than a crumbling façade and a big, empty lot—proportioned, like many traditional houses in Mérida's *centro histórico*, to extend deeply rearward from its narrow street frontage. Built flush with the sidewalk, the unassuming exterior bestows the dwelling with a seemingly humble aspect that belies its true nature, until it opens to reveal inner gardens, courtyards, and living spaces that stretch luxuriously into the distance.

The original stones of the façade hint at the building's origins, which date back to the seventeenth century. In adherence to the local conservancy's guidelines, Jorge was required to maintain the soaring 20-foot (6-meter) ceiling height in the original front structure, something he turned to his advantage by creating a grand entertaining area. Here, the artist's interventions add to the room's magic. Occupying center stage is a vast dining-room table he designed, inlaid with skulls at its corners, that comfortably seats sixteen. One wall is almost entirely covered by a mural that is a rendering of a Willem de Kooning painting. The geometric windows leading to the garden are draped with billowing lace fabric bought locally and layered to create surprising tones and iterations of color.

Beyond this first structure, which includes a catering kitchen and a wine cellar, Jorge was free to build as he wished. The result is a succession of three pavilions that merge with their environment and are divided by two jungle-like gardens of banana and ceiba trees, cacti, chili bushes, birds-of-paradise, and other native species. The central pavilion, where Jorge spends most of his time, contains the day-to-day kitchen and a living area. It is a large screened room that opens on both sides and has direct access to the pool. The ceiling height here is less dramatic than that of the entrance space, yet still optimal for showcasing Jorge's laser-cut pendant lamps that appear throughout the house, their shapes and colors inspired by tropical fruit such as papaya, and cocoa pods.

Nestled at the rear of the property, the third pavilion houses five bedrooms. Jorge skillfully transformed, in harmony with the rest of the house, the floors in the bedroom block into abstract paintings by employing zesty ceramic tiles procured from Guadalajara. Here, the lively greens and delicate blues of the tiles create a seamless dimension between the lush gardens and the dramatic skies.

BREAD

# XUNA'AN KAB

## *Mérida*

It was a hot and humid morning in Mérida when I arrived at the historical *barrio* of Ermita. Squinting from the harsh sunlight, I took note of the imposing yet subdued façade typical of the old houses along the Camino Real, a route that once connected Mérida with Campeche and received its name from the contributions made by the Spanish Crown for its construction in colonial times. I noticed a mass of overgrown vegetation peeking over the tall, austere wall and I suspected I was in for a treat.

Captivated by Yucatán's historic architecture and tropical climate, Bruce Bananto settled here in 2015 after living a hectic life as an architectural designer in New York. When he first laid eyes on the centuries-old property, it was a complete ruin, resembling a giant fortress with missing ceilings and enormous tropical trees growing inside. In a labor of love, Bruce painstakingly restored the house to its former glory. He was determined not to entrust its renovation to someone who didn't share his passion, and risk losing its character in the process.

I ventured with the owner throughout the house, admiring the monumental proportions and unusual murals. The inspiration behind the recurrent motifs covering the walls, designed and hand-painted by Bruce himself using ancient materials such as chalk paint, milk paint, tempera paint and wax paints, was the ceiba tree. Simply known as *la ceiba* in Spanish, or *ya'axche* in Maya, this tree is of huge cultural significance to the Maya civilization, both throughout history and today.

The architectural elements and furniture, all designed and largely handcrafted by Bruce, have a masculine yet whimsical component to them. The sprawling house is an eclectic mix of styles and influences, such as Tuscany, where he spent two years, as well as Moorish details inspired by Parroquia de Santiago Apóstol, one of Mérida's oldest churches. Turkish and Ottoman carpets and textiles were influential in the creation of the damask shapes on the walls that were based on historical motifs, but completely reimagined and manipulated by the designer and mixed with Elizabethan textile designs.

In addition to the unconventional mélange of styles, there is a distinct reference to nature throughout the house. *Xuna'an Kab*, the name given to it by the Maya-speaking stonemasons who worked with Bruce on its reconstruction, is a revered stingless honeybee species native to the rainforests of the Yucatán Peninsula. "I wanted to reflect the environment," he explained. "We are face to face with wild animals and tropical plants every day inside this city. I wanted to paint what I was experiencing."

JO BAER

# CASA VERSAILLES

## *Mérida*

It is a well-known fact that love affairs, in addition to bringing elation, a racing heart, and immense joy, can also induce feelings of consternation, frustration, and rage. Patricia Robert has felt all these emotions in relation to her house since their lives have been intertwined. Theirs was a love story that started when she first set eyes on the run-down dwelling in 2014.

Like so many people who end up in Mérida, Patricia did not mean to. Born and raised in Montreal, she first came to Yucatán chasing the sun and quickly became enchanted with its simple lifestyle and the Maya spirit. After thirteen years of frequenting the historic city with her husband Marcelo and son Émilio, and following a five-year-long search, they purchased their house in the Santiago neighborhood of the *centro histórico*, from someone who initially intended to restore it and live there but later had a change of heart. The house, with its charming French-style architecture, had clearly seen better days, despite having basic amenities like electricity and water. Neglected and in a sorry state of disrepair, it nevertheless entranced Patricia with its interesting L-shaped design and huge pink bougainvillea blooming wildly in the overgrown garden. Upon purchasing it, the seller handed over the plans he had commissioned for himself, but to Patricia's dismay, they were not what she imagined. Undeterred, she decided to forge her own path and create a fresh, new vision for their future home.

The restoration proved to be more challenging than she had anticipated. When asked why, she candidly admits, her fiery red hair shimmering under the relentless sun, "Well. It's Yucatán." Despite the hurdles, she was determined to preserve the essence of the house, respecting the original layout while extending the longer part of the L-shaped section to create space for two new bedrooms and bathrooms. Throughout the project, the building underwent significant reworking, with great care taken to restore the original doors and French glass, as well as the flooring in three of the rooms, which was relocated to the new extension.

During the rehabilitation process, Patricia found herself captivated by a collection of iron doors and windows that had captured her heart when she first acquired the old *casona*. Rather than discarding them, she decided to repurpose them creatively. Some were ingeniously transformed into a lattice, visually separating the master bedroom from its bathroom, while others were converted into a large desk that now stands proudly in her study.

With unwavering determination and an eye for one-of-a-kind details, Patricia brought her vision to life amidst the challenges posed by Yucatán's environment. Entering Patricia's world, all your senses are engaged at once. Her taste is wonderfully eclectic, and she has created a home that is feminine, joyful, and witty, and echoes her passion for Mexico.

N°5
CHANEL
CHANEL

# CASA DE LOS SANTOS

### *Izamal*

There is something about Izamal that makes one feel inextricably transported back in time. It's probably the languid clip-clop of the horse-drawn buggies making their away along the ancient cobblestones. It could be the Spanish colonial buildings and their striking yellow façades, or the manifold vestiges of Maya megaliths that dot the urban landscape. Known as The City of Three Cultures due to its rich mix of pre-Hispanic, colonial, and post-revolutionary history, the name Izamal means "dew that falls from the heavens" in Maya.

Fate had it that my first encounter with Casa de Los Santos was on a wet and stormy day. Wrapped in a veil of warm, drizzly rain, the eighteenth-century *casona* stood silently as a witness to Yucatán's turbulent past. Dusty *santos* and *vírgenes*, antique relics and dismembered religious figurines could be seen huddling inside, peering eerily into space through the tall windows that looked out onto the quiet street.

Nestled against the ancient ruins of the Maya pyramid Itzamatul, Casa de Los Santos was built in the eighteenth century, only steps away from the Convento de San Antonio de Padua. Having changed hands multiple times over its extensive history, the old *casona* is now owned by Coqui Coqui Perfumería, whose proprietors have transformed it into a perfume shop and a tranquil retreat for weary explorers. Its previous owner, Rodolfo Bolio, fondly known as Don Chulo, was an important and deeply religious member of the church, who often held prayer sessions in his centuries-old home. Little has changed since Don Chulo's industrious wife diligently polished the floors with Gas Morado, a popular brand of kerosene widely used in Yucatán for cleaning *pasta* tiles. On one of the peeling walls of the old dwelling, which still retains its domestic charm, hangs an enigmatic black-and-white photograph of the couple's smiling faces.

As I walked through the faded rooms, half intoxicated by the delicate scents of coconut, tobacco, and orange blossom wafting from the perfume bottles, I was transfixed by the abundance of spiritual references. Outside, the heavens ominously opened and the rain started pouring down.

# CASA SAN CRISTÓBAL

## *Mérida*

From the outside, Casa San Cristóbal looks like many of the one-storied homes in Mérida's historic center. With its tall, austere exterior and no way to peek inside, the only signs that something special lies beyond are the impeccably plastered wall and the two striking blue doors. These hints do not prepare me for the stylish oasis I encounter as I cross the threshold into the house that architectural designer Marc Perrotta shares with his husband, writer and editor John Newton. Their welcoming smiles as they usher me into their contemporary abode immediately make me feel at home.

A towering front room spans the width of the property, where on a concrete-surfaced table by local design firm Chuch Estudio, appear signs of a working life: with laptops, towers of books and messy piles of papers indicating that this is the couple's home office. At the far end of the room, above a low bookshelf that runs the length of the wall, is a large, unframed painting by Mérida artist Jorge Patrón Le Doux.

Through a tall opening, past a couple of bicycles perched casually in a corner, I enter the dining room, which is presided over by a round table from the turn of the century: "It was an engagement gift to Marc's parents, so he has sat at that table since he was born," says John. Both these spaces are the remainders of the house's original Spanish colonial-era structure. They form an L alongside a traditional courtyard with tall walls that frame the tropical sky above. Throughout the home, hand-plastered walls reveal the sweep of the arm that applied the texture, adding further warmth and patina to the interior, and connecting the new and historic sections.

I amble though the kitchen, making my way towards the back of the property, and go up a short flight of stairs that leads me into the living room. As the curtains are pulled back, I am pleasantly surprised to see a bright pink *casita*, partially hidden behind a lush jungle garden of heliconias, a ciricote tree, a banana tree, wild orchids, beach lilies, and plumerias. "The pink finish is a cement plaster with a touch of red additive. The color pink, specifically, provided a strong color that works well with the green of the jungle garden and the often blue skies of Mérida," explains Marc. A sense of peace engulfs me and I notice that the sounds of the street have faded away. "Here, you feel far removed from the city. It's a peaceful oasis where you hear birdsong more than traffic," says John.

# CASA DE MADERA

## *Izamal*

I vividly remember my first encounter with Casa de Madera. Steps away from the *plaza principal*, or main square, and nestled gracefully between similar one-storied houses, the small dwelling exuded an air of timeless solidity and natural beauty. Its weathered façade beckoned with a sense of intrigue, inviting me to step indoors and discover the secrets held within. Inside, tranquility reigned, providing a peaceful refuge from the bustling streets, and its wooden walls and simple decor created an atmosphere of rustic charm. As sunlight filtered through the windows, it cast a soft glow, revealing cozy spaces adorned with old furnishings and earthy accents.

Like many modest dwellings in Yucatán, the history of Casa de Madera, which means "wooden house", remains shrouded in mystery. While it is thought to have origins dating back to the seventeenth century, little is known of its past prior to the turn of the twentieth century. The house once functioned as a store for the workers of Hacienda Sacalá, which is located 3 miles (5 kilometers) north of Izamal. It provided essential goods during the flourishing era of the henequen industry. It is believed that the *trucs*—horse-drawn carts carrying bales of *sosquil*, or henequen fiber—would make stops there.

When the Hernández family, the current owners of Casa de Madera, found it, it was a mere ruin. However, as its sagging walls, made from planks of eastern white pine, were painstakingly cleaned away, the original stenciling began to emerge. This traditional artistic practice, deeply rooted in Yucatán's history, harks back to the colonial era, with influences from Spanish and indigenous artistic styles. Skilled artisans would carefully cut intricate patterns into thin sheets of metal or wood, creating templates adorned with geometric, floral, or symbolic motifs. These stencils were then pressed against the walls, and natural pigments like limestone-based paints were applied over them with care, transforming the walls into enchanting masterpieces.

The restoration of Casa de Madera proved to be a labor of love, carefully preserving its historical essence while breathing new life into its faded splendor. Master craftsmen meticulously repaired the wooden walls, ensuring the original character remained intact—and the addition of modern amenities is thoughtfully harmonious, blending the past with the present. Under the care of Private Haciendas, the charming old house stands gracefully, bearing the mark of time with aged walls that whisper stories of bygone eras.

CASA DE
MADERA

# CASA AZUL

## *Izamal*

One Saturday evening we drove into Izamal, our car windows rolled down and the hot wind in our hair. The dramatic blood-stained skies garnished with dashes of yellow seemed to reflect the ochre-colored façades of the colonial buildings of the *centro histórico*. While our eager eyes drank in the beauty of the history-laden streets, our car ventured over the cobblestones, snaking its way past the ancient temple of Kinich Kak Moo and towards our destination: Casa Azul.

A town of extraordinary beauty, with archeological vestiges speckling the urban landscape, Izamal was once an important hub for the Maya civilization, which reached its zenith here during the Classic Period, lasting from approximately 250 AD to 900 AD. When the Spanish conquistadors stumbled upon this wondrous land in the 1540s, it had been semi-abandoned for years, with only small clusters of dwellings scattered sparsely throughout the town. The first Spanish chronicles recount how the Maya, who had abandoned Izamal centuries earlier, would journey from afar to venerate their deities at this significant pilgrimage site. The Franciscans, who aimed to evangelize the indigenous people at this important location, also played a role in the acculturation and assimilation of their culture into Spanish colonial society by constructing the formidable Convento de San Antonio de Padua atop the pyramid of Pap Hol Chac in 1562.

Casa Azul is a charming, one-bedroom house steps away from the *plaza principal*, or the main square. If you lean out the front door and peer to the right, the omnipresent silhouette of the old convent dominates the landscape. Built during colonial times with stones from the neighboring pyramid of Itzamatul, the casita was purchased in 1999 by the Hernández family, who set about the restoration work with respect for the building's heritage.

Expert carpenters were brought in to salvage the wooden beams and as much of the original woodwork as possible, and yellow-tinted chukum was used for the new floors. Named after its main ingredient, chukum is a limestone-based stucco mixed with sap from the chukum tree, the colloquial name for *Havardia albicans*, a species endemic to Yucatán. An ancient technique lost with the decline of the classic Maya civilization, the creation of chukum was rediscovered towards the end of the 1990s by architect Salvador Reyes Ríos, who spent years experimenting with different mixes, recreating the time-tested recipe. The interior of the house was painted in a vibrant cobalt blue after multiple coats of old paint were scratched off the walls to reveal that, at a certain point in time, they had once been covered in that hue. The sparsely furnished rooms emphasizes the modesty of the dwelling, creating a monastic sense of calm.

# CASA DE LOS ARTISTAS

## *Izamal*

One early morning in February, I made my way across the main plaza of Izamal towards Casa de Los Artistas. The locals were going about their day, creating a vibrant and energetic scene; the scent of spicy street food wafted through the air, enticing my taste buds with the promise of delightful flavors. The golden-hued weathered walls of the colonial buildings surrounding the square added a lively splash of color to the bustling scene, and the intricate façades and graceful arches whispered secrets of the town's rich history. As I rounded the corner onto Calle 31, the main artery of Izamal, I encountered the unassuming exterior of Casa de Los Artistas.

Little is known about the history of this house and, like many colonial-era dwellings, the date of its construction remains a mystery. In 1945, Ricardo Reyes Bolio acquired the property and resided there with his family until 1967. Subsequently, Antonio Torres became the owner, passing it down to his daughter, María Guadalupe Torres Aranda, who eventually sold it to the Hernández family in 1998. Now a holiday-rental home managed by Private Haciendas, the house was originally conceived as a residence for artists from Fundación TAE (Transformation, Art, and Education) and underwent a significant restoration process in 2001, led by architect Salvador Reyes Ríos, that lasted for fourteen months.

The minimal interiors, thoughtfully designed by Reyes Ríos and Josefina Larraín, boast a vibrant array of colors. The color choices, which capture the lively essence of Yucatán, were influenced by two main inspirations. The first one was the enduring yellow hue of Izamal, which could still be seen on the tiles of the original corridor, despite their wear. This selection aimed to maintain a strong connection with the town upon entering the house, creating a chromatic unity that resonates with the essence of the location. Other colors and shades were derived from the layers of paint on the walls, ensuring a consistent presence of blues and reddish tones throughout the design. Most of the furniture, designed in collaboration with John Prentice Powell and Josue Ramos Espinosa, was custom-made in Mérida, with lighting, leather chairs, and mosaic tables crafted exclusively for each space. Other pieces, such as the wooden dining table, were sourced from antique shops.

I wandered from room to room and up a spiral staircase designed by Cuban American artist Jorge Pardo. As I reached the rooftop and took in the breathtaking views of Izamal, I was reminded of a model of the ancient town I had recently seen at El Gran Museo del Mundo Maya, featuring the prominent Convento de San Antonio de Padua and the Kinich Kak Moo pyramid. It included a solitary palm tree situated near the convent, precisely where Casa de Los Artistas stands today.

ACKNOWLEDGEMENTS

I am very deeply grateful to my husband, Pepe, to whom I owe so much. You are the calming presence that accompanies me on this wild adventure that is life. From day one you understood my enthusiasm and passion for Yucatán and were happy to accompany me as I traveled through the region unveiling its otherworldly treasures. Thank you always for your unwavering support.

My heartfelt thanks to my children, María and Pepe, my shining beacons of light. María, I will forever be grateful to you for helping me transform my dream into a reality. What would I have done without you? Pepe, I hope you know how much I appreciate your support and encouragement. You made the bleakest days possible and the brightest days delightful. To my granddaughter, little María, thank you for bringing such joy into my life. I hope this book will one day inspire you to always be curious and to lead a life of adventure. Your world is your oyster!

To my father and late mother, I deeply appreciate the unique upbringing you both provided me. It has proven to be an invaluable foundation for the existence I have chosen to lead. Your lives were shaped by your foreign travels, and you always encouraged me to go far and wide in my explorations. Mummy, my greatest role model, thank you for instilling in me an insatiable curiosity for life.

To my brother Jaime and my sister Becky, thank you for your wholehearted support. I appreciate your objective ears and eyes throughout this whole process.

*Grazie* to the brilliantly talented Guido Taroni. Your discerning eye and exquisite taste have been invaluable in bringing these beautiful houses and haciendas to life. I could not have dreamt of a better companion or a more talented photographer to do this book with. It has been an honor and a pleasure working with you.

A heartfelt thank you to my dear friend John Prentice Powell for always pointing me in the right direction. I am also deeply thankful for your kindness, and unlimited generosity and hospitality.

My appreciation to Luis Millet – thank you for so generously sharing your expertise and knowledge of the ancient Maya civilization with me.

Thank you to Joanna Roche and Leopoldo Sánchez for introducing me to Mérida for the first time and for always being so warm and welcoming.

Thank you to Marilú Hernández and Carola Díez for your inestimable advice and for transmitting your passion for Yucatán to me.

Thank you Martina Mondadori for your lovely foreword and for your unfaltering support.

I want to extend my sincere gratitude to Beatrice Vincenzini, my publisher, for believing in my vision. Your patience, dedication, and drive during the creation of *Inside Yucatán* have been truly invaluable. I also want to express my appreciation to Mark Magowan and the rest of the Vendome Press team, including Laura Capsoni, and Meredith Olson, my editor. Thank you all for your incredible contributions.

Last, but not least, I would like to thank all the homeowners who generously opened their doors to me, and in so doing, have made Yucatán and its magical private homes and haciendas better known to the rest of the world.

Susana Ordovás

INSIDE YUCATÁN
HIDDEN MÉRIDA AND BEYOND

First published in 2024 by The Vendome Press
Vendome is a registered trademark of The Vendome Press LLC

**www.vendomepress.com**

VENDOME PRESS US
PO Box 566
Palm Beach, FL 33480

VENDOME PRESS UK
132–134 Lots Road
London, SW10 0RJ



Distributed in North America by Abrams Books

ISBN 978-0-86565-445-7

PUBLISHERS
Beatrice Vincenzini, Mark Magowan, and Francesco Venturi

EDITOR
Meredith Olson

PRODUCTION DIRECTOR
Jim Spivey

PRODUCTION MANAGER
Amanda Mackie

DESIGNER
Laura Capsoni

Library of Congress Cataloging-in-Publication Data available upon request.

Printed and bound in China by RR Donnelley (Guangdong) Printing Solutions
First Printing